I Wonder Why the Sky Is Blue

Thelma Rea

New York

Published in 2002 by The Rosen Publishing Group, Inc.
29 East 21st Street, New York, NY 10010

First Library Edition 2002

Book Design: Haley Wilson

Photo Credits: Cover, pp. 1, 10 © Bruce Byers/FPG International; p. 4 © Gail Shumway/ FPG International; pp. 6, 16, 18 © Telegraph Colour Library/FPG International; p. 8 © Elizabeth Simpson/FPG International; p. 12 by Seth Dinnerman; p. 14 © Ken Ross/FPG International; p. 18 © Michael Simpson/FPG International; p. 20 © Walter Choroszewski/FPG International; p. 22 © Ron Chapple/FPG International.

Rea, Thelma, 1971-
I wonder why the sky is blue / Thelma Rea.
p. cm. — (The Rosen Publishing Group's reading room collection)
Includes index.
ISBN 0-8239-3724-0
1. Sky—Color—Juvenile literature. 2. Refraction—Juvenile literature. [1. Sky—Color. 2. Color. 3. Light.] I. Title. II. Series.
QC976.C6 R43 2002
551.56'6—dc21
2001007039

Manufactured in the United States of America

For More Information
Optics for Kids
http://www.opticalres.com/kidoptx.html

Contents

What Is the Sky?

When we are outside and look up, we see the sky. The sky is part of the **atmosphere**. The atmosphere is a **layer** of air that surrounds Earth. It is made up of different gases, water, and dust **particles**. Our moon does not have an atmosphere. The sky over the moon is always black.

The atmosphere helps change the color of the sky.

How Does Light Travel?

When sunlight shines on the surface of Earth, the light travels through the atmosphere in waves. Light waves are so small they cannot be seen. The light waves that move through the atmosphere make the colors that we see in the sky. Even moonlight comes from the sun. The moon **reflects** light from the sun.

Sunlight shines on the moon, which acts like a mirror and reflects the light.

How Are Colors Made?

Light from the sun is called white light. White light is a mix of all colors. When sunlight passes through a block of glass called a **prism**, the light is separated into all the colors of the rainbow. Color is made by different **wavelengths** of light. Each color has its own wavelength. Red has the longest wavelength. Violet and blue have the shortest wavelengths.

When light passes through a prism, we can see the separate colors of the rainbow.

Why Is the Sky Blue?

The color of the sky is caused by the way sunlight travels through the atmosphere. Sunlight is **scattered** by **trillions** of tiny particles in the air. The blue and violet wavelengths of light scatter more easily throughout the sky because they are so small. This creates the blue color you see in the sky during the day.

You can see many different shades of blue in the sky in one day.

How Does Light Scatter?

This experiment will show you what happens when light waves get broken up and scattered around in the air.

- Fill a glass with water.
- Stir in four drops of milk.
- Shine a flashlight against the side of the glass. The light will not shine through because the milk drops scatter the light. The particles in the atmosphere scatter sunlight in the same way.

Why Is the Ocean Blue?

The ocean looks blue because it reflects the color of the sky. When light shines on the top of the ocean, the water takes in all the wavelengths except for blue. The blue wavelengths are reflected, so we see the water as blue.

Oceans look blue because they reflect the color of the sky.

When Is the Sky Red?

The sky turns red or orange at sunset, when the sun is low in the sky. Light moves through the thickest layer of dust particles in the air during this time. Only the longest wavelengths of light, the reds and oranges, can move through the air, so those are the colors we see.

You can see many different colors in the sky during sunrises and sunsets.

What Other Colors Are in the Sky?

Colors change in the sky as light travels through the atmosphere. Different colors are created when light is scattered by particles and when the sun moves to a new place in the sky. The sky may turn purple or dark gray right before a storm.

The sky turns different colors when there are changes in the weather.

Why Are Clouds White?

We know that sunlight is really a mixture of all colors. We also know that the color of the sky is caused by tiny particles scattering more blue and violet wavelengths than any other color. When larger particles scatter light, they scatter the wavelengths of all colors. So when light hits a cloud, the bigger particles and water droplets in the cloud scatter all the light. This makes clouds look white.

Clouds can change from white to gray or black depending on how much water or dust they contain.

When Can You See Rainbows?

Rainbows often form right after a rainstorm when sunlight travels through raindrops that are still in the air. The raindrops are large enough to act as prisms. They split the sunlight into different colors. Rainbows are curved because raindrops are round. The raindrops bend the sunlight as it goes through them. When light bends, we see colors in a curved shape. Now when you look at the sky, you will know why you can see so many beautiful colors.

atmosphere	The layer of gases and dust that surrounds Earth.
layer	One thickness of something lying over or under another.
particle	A very small piece of something.
prism	A block of glass that separates white light into seven colors.
reflect	To give back light, heat, or sound.
scatter	To separate and go in different directions.
trillion	A large number that means one thousand billions.
wavelength	The amount of space between a point on one wave of light and a point in the same spot on the next wave.

Index